NATURE'S REVENGE

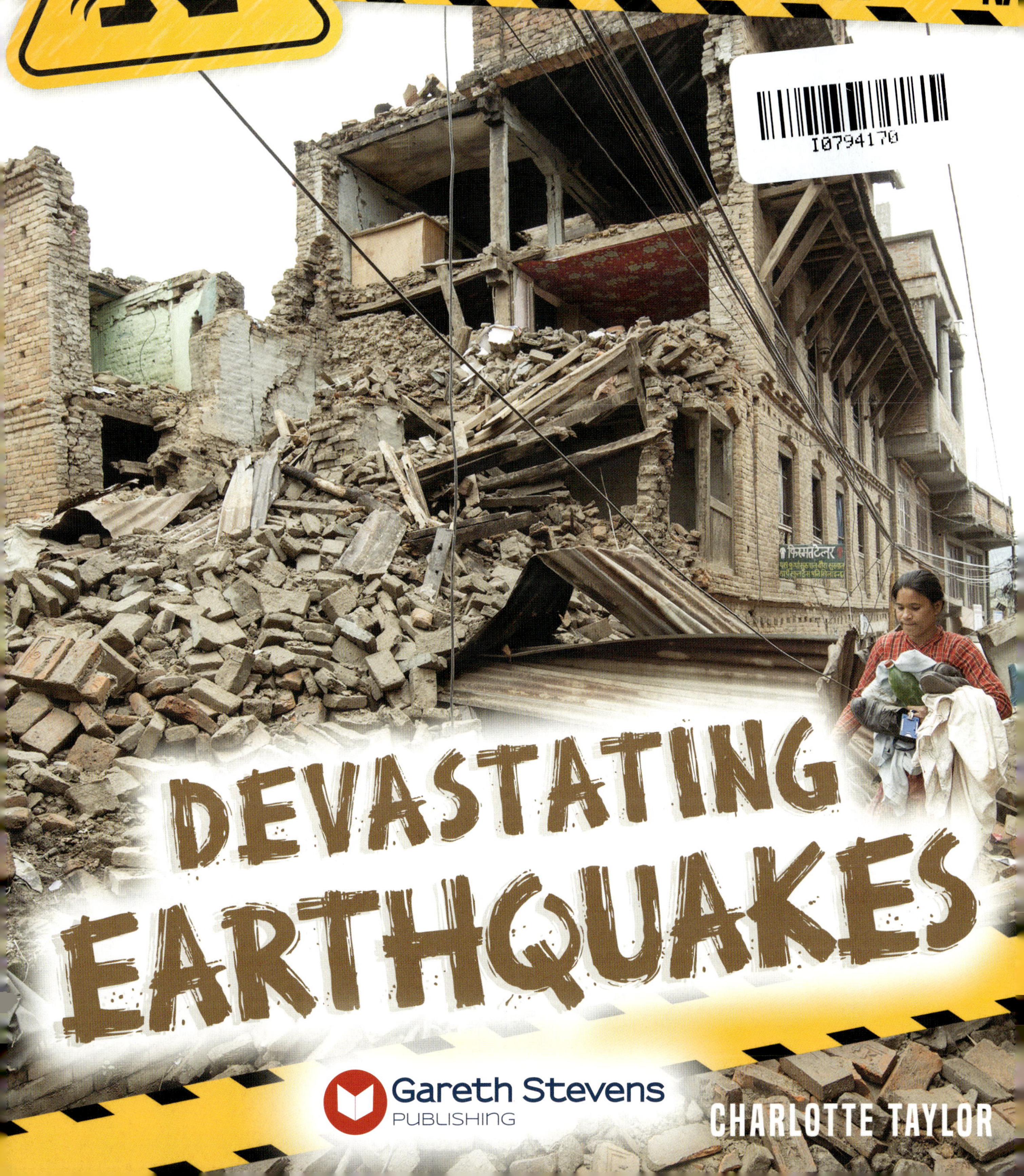

DEVASTATING EARTHQUAKES

Gareth Stevens PUBLISHING

CHARLOTTE TAYLOR

Please visit our website, www.garethstevens.com. For a free color catalog of all our high-quality books, call toll free 1-800-542-2595 or fax 1-877-542-2596.

Cataloging-in-Publication Data
Names: Taylor, Charlotte, 1978-.
Title: Devastating earthquakes / Charlotte Taylor.
Description: New York : Gareth Stevens Publishing, 2023. | Series: Nature's revenge | Includes glossary and index.
Identifiers: ISBN 9781538280515 (pbk.) | ISBN 9781538280539 (library bound) | ISBN 9781538280522 (6pack) | ISBN 9781538280546 (ebook)
Subjects: LCSH: Earthquakes–Juvenile literature. | Natural disasters–Juvenile literature.
Classification: LCC QE521.3 T39 2023 | DDC 551.22–dc23

Portions of this work were originally authored by Therese M. Shea and published as *Rocked by Earthquakes*. All new material in this edition was authored by Charlotte Taylor.

Published in 2023 by
Gareth Stevens Publishing
29 East 21st Street
New York, NY 10010

Designer: Leslie Taylor
Editor: Megan Quick

Photo credits: Cover (Nepal rubble), p. 5 think4photop/Shutterstock.com; pp. 1–32 (icons-series artwork) Vector by/Shutterstock.com; pp. 1–32 (hazard tape-series artwork) DDevecee/Shutterstock.com; p. 7 Sakurra/Shutterstock.com; p. 9 Designua/Shutterstock.com; p. 11 (wave diagram) VectorMine/Shutterstock.com; p. 11 (aftermath photo) mTaira/Shutterstock.com; p. 13 (Albania) Ajdin Kamber/Shutterstock.com; p. 15 (seismologist) Mila Supinskaya Glashchenko/Shutterstock.com; p. 16 (diagram) trgrowth/Shutterstock.com; p. 17 (bridge) https://commons.wikimedia.org/wiki/File:Miles_Glacier_Bridge,_damage_and_kludge,_1984.jpg; p. 19 Everett Collection Historical/Alamy.com; p. 20 (flooded cars) https://commons.wikimedia.org/wiki/File:Sendai_Airport_after_the_tsunami.jpg; p. 21 https://commons.wikimedia.org/wiki/File:US_Navy_110315-N-5503T-307_An_aerial_view_of_damage_to_Otsuchi,_Japan,_after_a_9.0_magnitude_earthquake_and_subsequent_tsunami_devastated_the_area_in_northern_Japan.jpg; p. 23 (rescue dogs) 4.murat/Shutterstock.com; p. 25 (graphic) mark joseph mabuna/Shutterstock; p.25 (emergency kit) Pixsooz/Shutterstock.com; p. 27 https://commons.wikimedia.org/wiki/File:2018_Earthquake_In_Hualien_City_Taiwan_(245591057).jpeg; p. 29 (volunteers) Mihai Speteanu/Shutterstock.com.

Printed in the United States of America

CPSIA compliance information: Batch #CSGS23: For further information contact Gareth Stevens, New York, New York at 1-800-542-2595.

CONTENTS

On Shaky Ground 4
A Look Inside Earth 6
Bumping Plates 8
Seismic Waves 10
Measuring Quakes 14
Danger and Destruction 18
Help on the Way 22
Be Prepared 24
Stronger and Safer 26
Hope for the Future 28
Glossary 30
For More Information 31
Index 32

Words in the glossary appear in **bold** type the first time they are used in the text.

ON SHAKY GROUND

Big earthquakes don't happen too often, but when they do—look out! The ground starts to shake. It's hard to stand up. Objects may fall over or be knocked off shelves. The shaking can last a few seconds or a few minutes, but that's all it takes for an earthquake to cause a lot of **damage**.

Major earthquakes are uncommon, but smaller ones happen all the time. They cause little or no damage. When a large earthquake does happen, it can be **destructive** and deadly, especially since these events cannot be **predicted**. Luckily, scientists are learning more about why and where earthquakes happen so people can be better prepared.

Two major earthquakes hit Nepal in 2015, destroying many buildings and killing thousands of people.

Extreme Nature

When a big earthquake hits an area where people live, it is called a natural disaster. A natural disaster is an event in nature that causes lots of damage. These **extreme** events can destroy buildings, knock out power, and hurt or kill people and animals. Earthquakes are just one kind of natural disaster. Others include floods, **hurricanes**, and blizzards.

A LOOK INSIDE EARTH

If you want to understand earthquakes, start by taking a look at what's inside Earth. Our planet is made up of different layers. At the center of Earth is the core, which is super hot. The next layer is called the mantle, and the outer layer is the crust. Earth's crust and the top part of the mantle are known as the lithosphere.

The lithosphere is made up of 15 to 20 moving pieces, called tectonic plates. These rocky plates fit together a lot like puzzle pieces, but with a little space in between. This allows them to move around on top of the hot, soft rock in the mantle.

Earth on the Move

Tectonic plates move at different speeds, but none of them move fast! On average, the plates move about 4 inches (10 cm) in a year. Even though this is very slow, the plates have been moving for a very long time. This means they have moved about 62.5 miles (100 km) in the past million years.

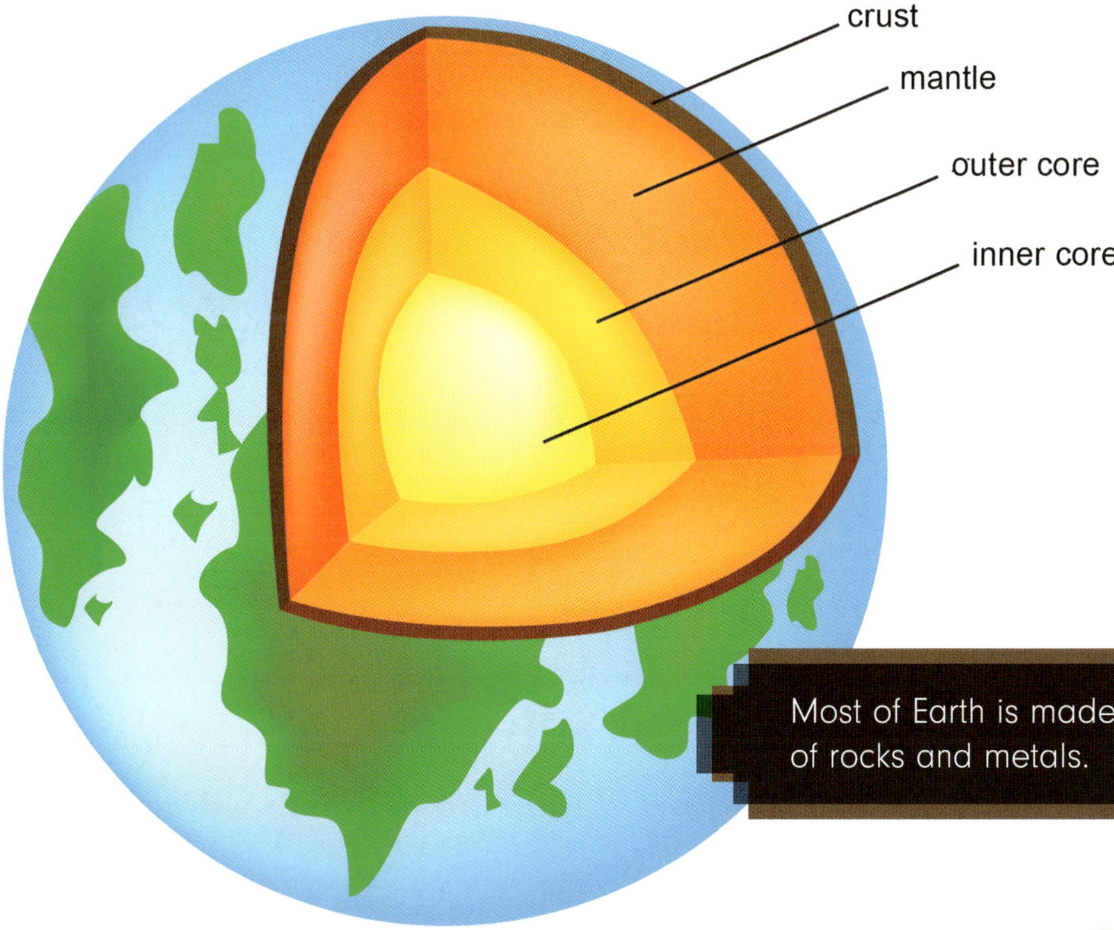

Most of Earth is made of different type of rocks and metals.

BUMPING PLATES

As tectonic plates move along, they bump into one another. The edges of the plates, called plate boundaries, may pull away or push towards each other. One plate may move over or under the other. And sometimes, one piece slides past the other. But the edges of the plates are uneven, so the rocky pieces may get stuck. This is where trouble can start.

When plates get stuck, they try to keep moving. Pressure builds up. Finally, some of the rocks in the plates break, and the plates slip past each other. The pressure releases and an earthquake occurs. The area where the broken rocks move past each other is a fault. The area above the fault, on Earth's surface, is the epicenter.

TECTONIC PLATES

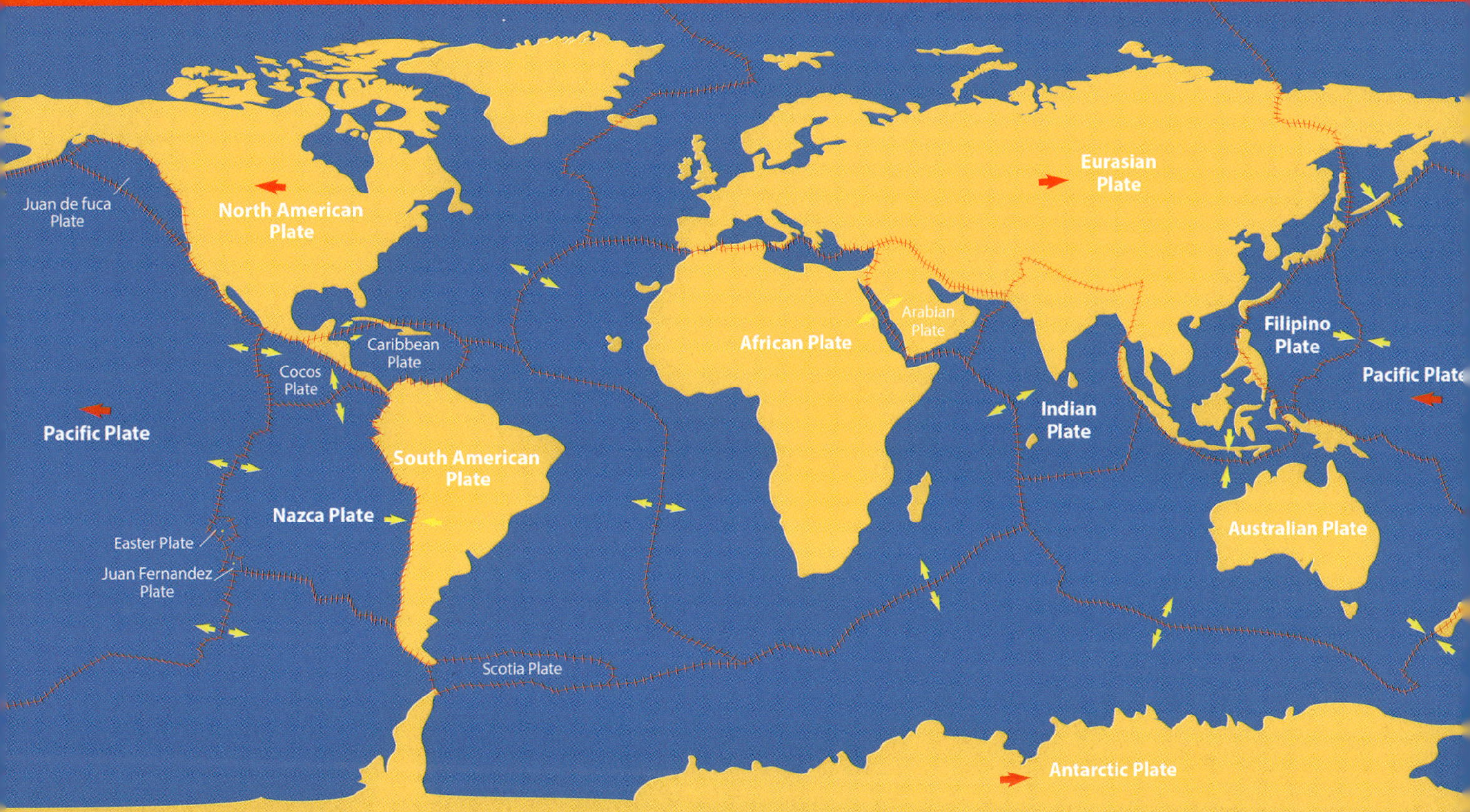

The red lines on this world map show the boundaries of each tectonic plate. As you can see, some plates are much larger than others.

Where's the Fault?

Most earthquakes happen along plate boundaries, where there are lots of faults. But sometimes a fault occurs away from the plate edge. Within a single plate, pieces can move in different directions. This means earthquakes can happen anywhere. It also makes it harder to know where a future earthquake might take place.

SEISMIC WAVES

When an earthquake occurs, a huge amount of pressure is released. This pressure moves outward as waves of energy. These are called seismic waves. The waves travel through Earth, shaking the ground and everything connected to it. The most powerful earthquakes cause seismic waves that can be felt thousands of miles from the epicenter.

Seismic waves that travel through the inner Earth are called body waves. P, or primary, body waves squeeze and expand rock as they travel. S, or secondary, body waves move up and down or side to side. S waves are slower than P waves.

A huge earthquake hit near Honshu, Japan, in 2011. The quake's seismic waves could be felt as far away as Russia.

Shake It Up

Since each body wave moves differently, each one has a different effect when it comes to Earth's surface. When a strong P wave shoots up to the surface, it might push a building up and down. But when an S wave comes from the same direction, the building shakes from side to side.

Once body waves move through Earth, a different and more destructive type of wave can follow. Surface waves get their name because they travel along the surface of Earth. They do not move as fast as body waves, but they are bigger. They cause the damage we see at ground level.

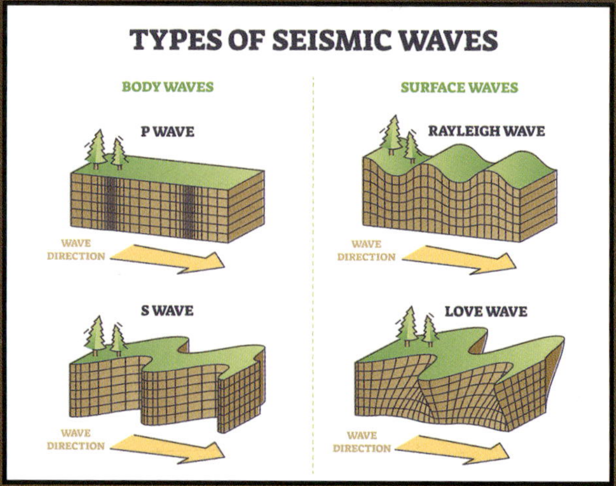

There are two types of surface waves that are named after the scientists who first identified them. Love waves move from side to side. Rayleigh waves move like a wave on the water, up and down and side to side. Of the two, Love waves move faster.

Setting Waves in Motion

Shifting tectonic plates are not the only cause of seismic waves. Volcanoes are another possible source. The liquid or partly liquid rock that moves underneath a volcano can make changes in the crust and cause seismic waves. Sometimes, people are to blame. Underground **mines**, explosions, and large dams are a few ways human activity has set off seismic waves.

A 2019 earthquake in Albania was the strongest to hit the area in many years. Surface waves destroyed many buildings like this one.

MEASURING QUAKES

A seismologist is a scientist who studies earthquakes. Seismologists use instruments called seismographs to measure movement in the ground during an earthquake. A simple seismograph has a base in the ground and a weight hanging from a spring. This weight is called a pendulum. When the ground shakes, the base shakes and the seismograph records the movement, sometimes on paper. That record of an earthquake is called a seismogram.

In a seismogram, a line with few waves means an earthquake was small, but a very wavy line means it was large. P waves are recorded first, followed by S waves, and then surface waves. Scientists use the waves of a seismogram to figure out the strength of an earthquake.

The change in wave size on a seismogram like this one shows scientists when the strongest seismic waves began in an earthquake.

A Career in Earthquakes

Would you like to study earthquakes as your job? There are lots of ways to do this. Seismologists focus on seismic waves to understand why and how earthquakes occur. Geologists study rocks and faults and what happens in Earth before, during, and after a quake. Other job possibilities are teaching, **emergency** services, or working for oil or mining companies.

Once scientists know how strong an earthquake was, they give it a number. The most famous earthquake measurement scale is called the Richter **magnitude** scale. Scientists use the largest seismic wave on a seismogram to figure out the earthquake's magnitude. Earthquakes with a magnitude of 2.0 or less aren't usually felt by people. Those with magnitudes of 4.0 or less often don't cause much damage. A magnitude of 6.0 or higher often causes major damage.

The Richter scale is not the only way to measure a quake's magnitude. In fact, many scientists today prefer to use the moment magnitude scale, which they think measures an earthquake's strength better. Unlike the Richter scale, a moment magnitude measurement reflects the total amount of energy released at a fault.

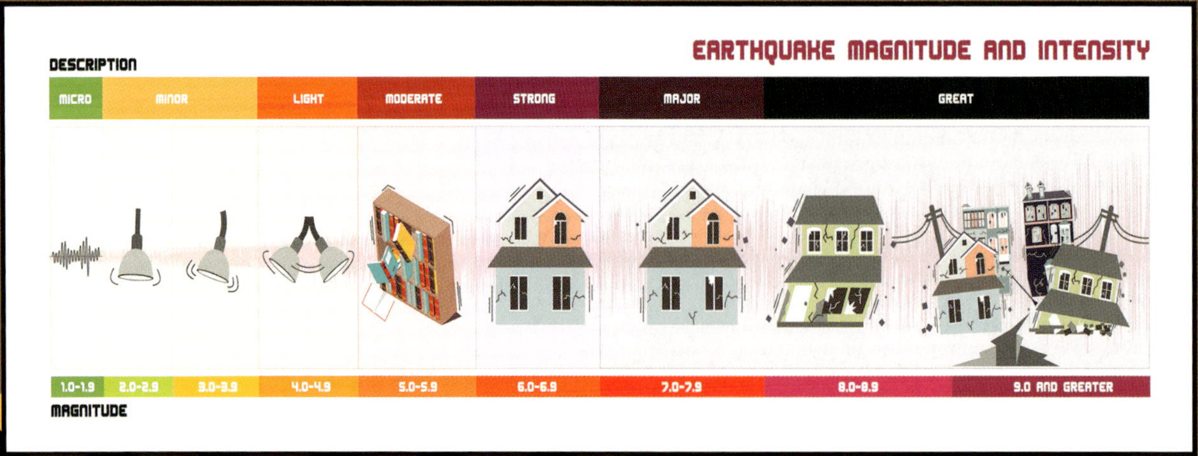

The Great Alaskan Earthquake of 1964 caused the ground to shift and damaged the Miles Glacier Bridge. A short-term fix to the bridge is shown here.

The Great Alaskan Earthquake

The strongest earthquake to strike the United States occurred on the coast of Alaska in 1964. It measured an unbelievable 9.2 on the Richter scale. It was also the second strongest earthquake ever recorded in the world! The quake triggered a **tsunami** and was felt all the way to California. The earthquake and tsunami killed 131 people.

DANGER AND DESTRUCTION

An earthquake's location is a huge part of how much damage it causes. If a strong quake happens in the middle of a large desert with no buildings, it won't do much harm. But if one strikes a big city, that means trouble. When buildings and bridges cannot handle the shaking, they sink into softened ground or fall. People in and on these structures can be hurt or killed.

The moving ground can also crack roads and railways, causing cars, trucks, and trains to turn off course. Another danger is the breaking of dams and levees, which hold back large amounts of water. The release of this water can cause flooding, property damage, and death by drowning.

The Great San Francisco Earthquake of 1906 caused three days of fires around the city.

Only in the Movies

Many dangers go along with an earthquake, but they do not include falling into a giant hole in Earth! This may happen in the movies or on TV, but in real life it's not possible. The ground around a fault does not pull apart—the two sides just push past each other. The ground may crack or shift, but Earth does not open up.

Earthquakes can also cause damage in nature. They can cause landslides and mudslides in hilly areas with loose ground. These can be very dangerous, even burying buildings and people. And earthquakes under the ocean can cause giant waves called tsunamis. When the waves reach shore, they can knock down buildings, flood towns, and drown people.

The dangers of earthquakes are not limited to the quake itself. Shaking movements before or after an earthquake are called tremors. When they happen before the quake, they are called foreshocks, and those that come after are aftershocks. They may occur in the days, weeks, or months after a major earthquake.

Little Earthquakes

An aftershock is a bit like a mini-earthquake that happens after the mainshock, or main earthquake. In an aftershock, the land around the fault continues to shift. Aftershocks become less frequent over time. Sometimes an aftershock is stronger than the mainshock. When this happens, the mainshock is renamed a foreshock, and the aftershock becomes the mainshock.

A 2011 tsunami in Japan was triggered by a magnitude 9.0 earthquake under the ocean floor. Almost 20,000 people died. A closeup of a flooded airport is shown on page 20.

HELP ON THE WAY

Once an earthquake hits an area, lots of people jump into action. Getting help quickly can save lives when people are in danger. Local, state, national, and sometimes **international** teams, including fire, medical, and rescue workers, are all part of emergency operations after an earthquake.

Teams of workers look for survivors who may be trapped inside fallen buildings. They set up places where people can get medical help if they need it. They help people find their missing loved ones. They may set up shelters for people who have lost their homes. People also need clean water and food in the days after an earthquake hits.

Rescue workers use rescue dogs to help search for survivors after a large earthquake in Turkey in 2011.

Furry Heroes

Not all emergency workers walk on two legs! Dogs can be an important part of a search and rescue team. They can use their strong sense of smell to find a person buried under debris and they can reach small spaces that humans cannot. Trained rescue dogs travel around the world to help out after a disaster like an earthquake.

BE PREPARED

When people hear that a storm is on the way, they usually have some time to prepare. But scientists can't predict earthquakes, so it's important to be ready now. Have emergency supplies on hand: a first-aid kit, a battery-powered radio, a cell phone charger, blankets, bottled water, canned food, and a flashlight with extra batteries. A kit like this is helpful in many types of emergencies.

To prepare your house for an earthquake, make sure there are no heavy objects on shelves. These and other pieces of furniture can be **anchored** to walls and floors so they're less likely to move. You and your family should also talk about a good place to meet in case you're separated during an earthquake.

International ShakeOut Day, held on the third Thursday in October, is when people around the world practice earthquake safety.

Know What to Do

If an earthquake strikes, try not to panic. If you're indoors, stay near the center of the building near a wall, where the structure is strongest. Get on your hands and knees and cover your head and neck. You could also crawl under a sturdy table. Don't stay near windows and doors or near things that could fall on you, such as lights.

STRONGER AND SAFER

Most buildings made in the United States today have to follow rules that ensure they will be as safe as possible if an earthquake hits. Engineers and **architects** plan buildings to be **symmetrical** so no part of the structure is weaker than another. They don't include decorations that could easily fall and hurt people.

In high-risk areas, engineers use materials with high ductility—they can stretch or change shape without breaking. Materials like steel have high ductility, allowing buildings and bridges to bend. Within a building, special systems are placed in the floors and roof. When the building experiences a sideways force, the systems work with vertical structures to withstand it. For extra strength, engineers use diagonal pieces called trusses in frames, sometimes crossing them to make an "X" shape.

A 2018 earthquake in Taiwan caused this building to partly collapse, killing 14 people inside. Several people were jailed for not following safety rules when the building was made.

West Coast Earthquakes

California is known for its earthquakes, thanks to its location on the large San Andreas Fault. But Alaska tops the list for most quakes in the United States. Alaska has a great deal of open space. Earthquakes there often occur where there are few buildings or people. California has a large population with big cities and buildings, so an earthquake is more likely to cause major damage.

HOPE FOR THE FUTURE

Our Earth is constantly shifting and shaking. Most of the time, no one feels these small movements. Once in a while, a major event happens that can cause damage and death. With no way to tell when and where an earthquake will hit next, the smartest move is to be prepared. This is especially true if you live in a high-risk area such as the west coast of the United States.

While we cannot predict earthquakes right now, this might not always be the case. Scientists are developing new tools and learning more about earthquakes all the time. They hope that one day they will be able to predict an earthquake and save many lives.

More Earthquakes?

Earth is getting hotter, partly because of everyday human activities like driving cars and heating or cooling homes. This is called global warming. It's causing **glacier** ice to melt and oceans to rise. Some people believe rising sea levels will cause more stress on Earth and more earthquakes. More research needs to be done before we know. We do know for sure that global warming is causing more natural disasters like hurricanes, floods, and blizzards.

After a natural disaster, people often come together to help each other. These volunteers are helping to rebuild a school that was destroyed in an earthquake in Nepal.

GLOSSARY

anchor: to fix or secure

architect: a person who designs buildings

damage: harm

destructive: causing something to be destroyed

emergency: an unexpected situation that needs quick action

extreme: great or severe

glacier: a very large area of ice that moves slowly over a wide area of land

hurricane: a powerful storm that forms over water and causes heavy rainfall and high winds

international: involving two or more countries

magnitude: a measure of the power of an earthquake

mine: a pit or tunnel containing rocks and other matter

predict: to guess what will happen in the future based on facts or knowledge

symmetrical: having an equal balance of something on both sides of a dividing line or around a center

tsunami: a huge wave of water created by an underwater earthquake or volcano

FOR MORE INFORMATION

BOOKS

Bybee, Veeda. *Lily and the Great Quake: A San Francisco Earthquake Survival Story.* North Mankato, MN: Stone Arch Books, 2020.

Koontz, Robin. *Disaster-Proof!* Vero Beach, FL: Rourke Educational Media, 2017.

Romero, Libby. *All About Earthquakes.* New York, NY: Children's Press, 2021.

WEBSITES

NASA Climate Kids: Weather and Climate
climatekids.nasa.gov/menu/weather-and-climate/
Check out articles, videos, and photos about our changing planet.

US Geological Survey: Earthquakes for Kids
earthquake.usgs.gov/learn/kids/
This site includes earthquake history, photos and videos, and ideas for science projects.

Weather Wiz Kids: Earthquakes
www.weatherwizkids.com/weather-earthquake.htm
Learn more about earthquakes with pictures, videos, and fun facts.

INDEX

aftershock 20, 21

body waves 10, 11, 12, 14

core 6, 7

crust 6, 7, 12

ductility 26

epicenter 8, 10

fault 8, 9, 15, 16, 19, 21, 27

foreshock 20, 21

geologist 15

global warming 29

Great Alaska Earthquake 17

Great San Francisco Earthquake 19

lithosphere 6

Love waves 12

mantle 6, 7

moment magnitude scale 16

pendulum 14

Rayleigh waves 12

Richter magnitude scale 16, 17

San Andreas Fault 27

seismogram 14, 15, 16

seismograph 14

seismologist 14, 15

surface waves 12, 14

tectonic plates 6, 7, 8, 9, 12

tremor 20

truss 26

tsunami 17, 20, 21